SCIENTIFIC AMERICAN EDUCATIONAL PUBLISHING

SCIENTIFIC AMERICAN INVESTIGATES SPIDERS

BROWN RECLUSE SPIDERS

BY ROXANNE ALDOLPHO

Published in 2025 by The Rosen Publishing Group
in association with Scientific American Educational Publishing
2544 Clinton Street, Buffalo NY 14224

Library of Congress Cataloging-in-Publication Data
Names: Aldolpho, Roxanne, author.
Title: Brown recluse spiders / Roxanne Aldolpho.
Description: [Buffalo, NY] : Scientific American Educational Publishing, an imprint of Rosen Publishing, [2025] | Series: Scientific American investigates spiders | Includes bibliographical references and index.
Identifiers: LCCN 2023059162 | ISBN 9781725351219 (library binding) | ISBN 9781725351202 (paperback) | ISBN 9781725351226 (ebook)
Subjects: LCSH: Brown recluse spider–Juvenile literature. | Spiders–Juvenile literature.
Classification: LCC QL458.42.L6 A43 2025 | DDC 595.4/4–dc23/eng/20240125
LC record available at https://lccn.loc.gov/2023059162

Portions of this work were originally authored by Sarah Machajewski and published as *Brown Recluse Spiders*. All new material in this edition is authored by Roxanne Aldolpho.

Designer: Andrea Davison-Bartolotta
Editor: Brianna Propis

Photo credits: Cover, pp. 1, 5, 15 Pong Wira/Shutterstock.com; p. 7 memv23/Shutterstock.com; p. 9 Nick626/Shutterstock.com; p. 11 (bottom) Edvard Mizsei/Shutterstock.com; p. 11 (map) Mister_X/Shutterstock.com; p. 11 (cartoon spider) Maquiladora/Shutterstock.com; p. 13 Steve Collender/Shutterstock.com; p. 17 Macronatura.es/Shutterstock.com; p. 19 Nick626/Shutterstock.com; p. 21 (web) Chinch/Shutterstock.com; p. 21 (main) Breck P. Kent/Shutterstock.com.

Printed in the United States of America

CPSIA compliance information: Batch #CSSA25. For Further Information contact Rosen Publishing at 1-800-237-9932.

CONTENTS

Words in the glossary appear in **bold** type the first time they are used in the text.

SMALL, YET SCARY

If you ever find yourself going through a box in your attic, looking underneath your furniture, deep in your closet, or in a corner of your basement, be wary. You may find yourself face-to-face with a brown recluse spider!

Brown recluse spiders are feared by many. These eight-legged terrors are small but have a bad **reputation**. Their bite can be very serious. However, they're very shy and probably more afraid of us than we are of them! Let's learn all about this fascinating creature.

Brown recluse spiders are excellent at hiding!

ARACHNIDS

More than 50,000 species, or kinds, of spiders are living in the world today. The brown recluse is just one species, but it's widely known because of its **venomous** bite.

Spiders are part of an animal group called arachnids. Tarantulas, scorpions, ticks, and mites belong to this group too. It's safe to say that arachnids aren't the most popular animals around—especially spiders. Even though many of them are harmless, even the tiniest spiders can make many people scream!

Brown recluse spiders are arachnids. Arachnids are a group of animals with segmented bodies, or bodies divided into two parts, and eight legs.

FUN FACT

THE LARGEST SPECIES OF SPIDER IN THE WORLD IS THE GOLIATH BIRD-EATER. THEY ARE A TYPE OF TARANTULA AND CAN GROW UP TO 11 INCHES (27 CM)!

SIX EYES AND A VIOLIN

Brown recluse spiders are brown and creepy, but they're not very large. These creatures are about 3/8 inch (1 cm) long and 3/16 inch (0.5 cm) wide.

Unlike other spiders, the brown recluse has six eyes. Most other types of spiders have eight. The recluse's eyes are arranged in three rows across its head. Also, on its head are **distinctive** markings that look like a violin. The violin shape can be hard to see, but the eyes don't lie. If you count six eyes on a spider, then it's definitely a recluse!

Can you spot the violin-shaped mark on this recluse?

FUN FACT

ONLY ABOUT 100 KINDS OF SPIDERS HAVE VENOM THAT CAN HARM HUMANS. TWO OF THEM LIVE IN THE UNITED STATES—THE BLACK WIDOW AND THE BROWN RECLUSE!

WHERE COULD YOU SEE ONE?

Brown spiders live in many different places. When you see one, you might be scared it's a brown recluse. But while this species is the most common of brown spiders, it doesn't live all over. Its habitat, or where it lives, is mostly in the southern and central United States.

They prefer to live in caves, under rocks, and in log piles. These spiders can also travel from one place to another if they're hiding in something, such as a moving box.

FUN FACT

BROWN RECLUSE SPIDERS ARE EXTREMELY ADAPTABLE AND CAN SURVIVE IN TEMPERATURES AS LOW AS 45°F (7°C) AND AS HIGH AS 110°F (43°C).

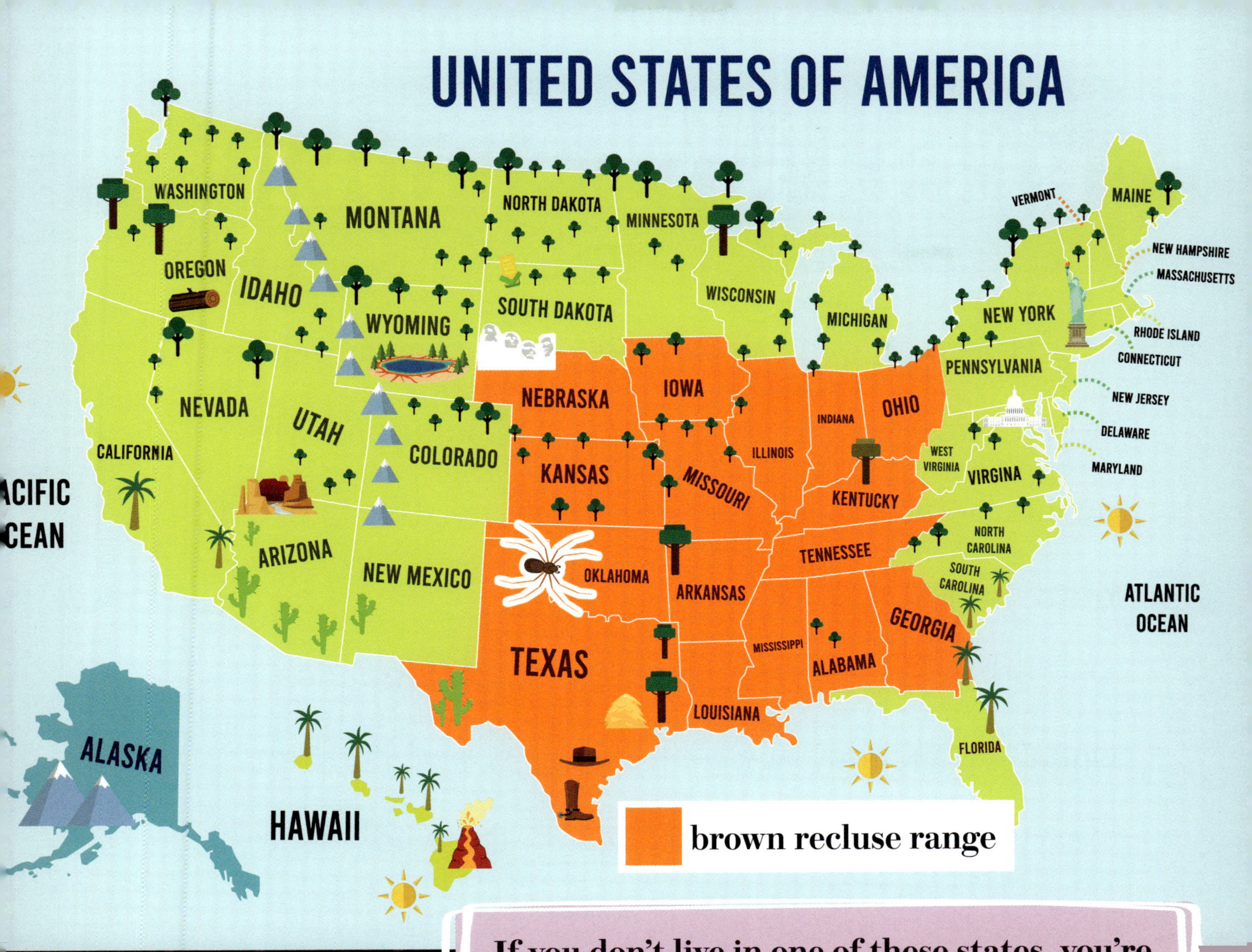

If you don't live in one of these states, you're much less likely to spot a brown recluse!

RECLUSIVE SPIDERS

Have you ever wondered where brown recluse spiders got their name? A recluse is someone who avoids people, preferring to live alone. This is exactly what the spider does!

Brown recluse spiders are solitary creatures, which means they live alone. They spend much of their time hiding in dark places. In the wild, these spiders live under logs and rocks. However, you might find them hiding under cardboard boxes, garbage cans, old tires, and even inside your shoe!

Some people have even found brown recluses hiding in their bed sheets. Yuck!

FUN FACT

BROWN RECLUSE SPIDERS ARE NOCTURNAL, MEANING THEY'RE ACTIVE AT NIGHT AND HIDE DURING THE DAY. THIS IS WHY YOU PROBABLY WOULDN'T CATCH THEM SNEAKING INSIDE YOUR BOXES OR SHOES!

PROWLING

Many spiders build silky webs to catch their meals. Spider webs are commonly very strong so they can be good traps for bugs. Most web-weaving spiders wait in their web until their **prey** comes along, but the brown recluse catches prey a bit differently.

The brown recluse spider doesn't use its web to catch food. Instead, it's a hunter, prowling for and preying on bugs both dead and alive. The recluse has venomous **fangs** that release toxic venom in their bite.

Brown recluse spider webs are loose, irregular, and off-white in color. They're built low to the ground or in corners. They use their webs for nesting rather than hunting.

FUN FACT

THE BROWN RECLUSE DOESN'T HAVE TO EAT OFTEN TO STAY ALIVE. IT CAN GO WITHOUT FOOD FOR MONTHS, AND IT HAS EVEN BEEN REPORTED TO GO TWO YEARS WITHOUT EATING!

BABY SPIDERS

Brown recluse spiders prefer to be alone, but that doesn't mean you can't find a group living together or near each other. **Infestations** happen quickly because female spiders lay 40 to 50 eggs at a time!

Female spiders **weave** an egg sac into their web. The tiny spiders that hatch, or come out, are called spiderlings. They slowly get bigger over time as they molt, or lose their skin. Molting occurs between five and eight times until the spiders become adults.

Brown recluse spiderlings turn a darker shade of brown as they age.

FUN FACT

BROWN RECLUSES HAVE LONG LIFE SPANS FOR SPIDERS! THEY **MATURE** AFTER A YEAR AND CAN LIVE TWO TO FOUR YEARS.

WHAT HAPPENS IF YOU'RE BITTEN?

The biggest reason brown recluses are feared is because of their venomous bite. The spider bites when it feels **threatened**, such as when you put on a shoe it's chosen to nest in or if you roll over one that's hiding in your bed.

People don't commonly feel the bite when it happens, though. **Symptoms** take hours to set in. The bite soon becomes red, swollen, and sore. A **blister** may appear, and some people start to feel pain and itchiness around the bite.

A brown recluse spider's bite can be serious. It's important to call your doctor immediately if you're bitten!

A BAD REPUTATION

It's easy to see why brown recluse spiders have a bad reputation. Their bite is venomous, and they hide and creep around where we can't see them.

These facts make the brown recluse sound like a true terror, but—believe it or not—they're not as bad as they seem. Brown recluses hunt bugs, which helps to control the pest population. And they don't seek people out. But if you see one, watch out. They are still willing to bite!

Ouch!

A BROWN RECLUSE BIT ME! NOW WHAT?

SIGNS:

- red, itchy skin
- blister or sore with irregular edges
- skin turning black or purple

SYMPTOMS:

- **fever**
- chills
- throwing up
- headache
- itching

NEXT STEPS:

- If you think you've been bitten, tell an adult right away.
- Put ice on the bite.
- Get to the doctor right away!

GLOSSARY

blister: A painful swelling on the skin.

distinctive: Having to do with a feature of something that sets it apart from others.

fang: A hard, sharp-pointed body part a spider uses to put venom into its prey.

fever: A raised body temperature due to sickness.

infestation: An unusually large number of bugs or animals in one place.

mature: Fully grown.

prey: An animal that's hunted by other animals for food.

reputation: The views that are held about something or someone.

symptom: A sign that shows someone is sick.

threatened: Feeling like danger is coming.

venomous: Able to inject venom, a poison made in the body of some animals that can cause harm.

weave: To make by twining threads together.

FOR MORE INFORMATION

Books

Gibbons, Gail. *Spiders*. New York, NY: Holiday House, 2020.

Higgins, Melissa. *Splendid Spiders*. North Mankato, MN: Pebble, 2020.

Websites

Britannica Kids: Brown Recluse
https://kids.britannica.com/students/article/brown-recluse/
Learn more about the habits and lifestyle of the brown recluse.

Brown Recluse Spider Bite
https://www.columbiadoctors.org/health-library/condition/brown-recluse-spider-bite/
You can never be too careful—read about what to do if you're ever bitten by a brown recluse.

Publisher's note to educators and parents: Our editors have carefully reviewed these websites to ensure that they are suitable for students. Many websites change frequently, however, and we cannot guarantee that a site's future contents will continue to meet our high standards of quality and educational value. Be advised that students should be closely supervised whenever they access the internet.

INDEX